Learning
About Money
Grades 1–2

by
Robyn Silbey

Cover Illustration by
Marlene Albright

Published by Frank Schaffer Publications
an imprint of

Children's Publishing

Author: Robyn Silbey
Cover Illustration: Marlene Albright

Children's Publishing

Published by Frank Schaffer Publications
An imprint of McGraw-Hill Children's Publishing
Copyright © 1985 McGraw-Hill Children's Publishing

Send all inquiries to:
McGraw-Hill Children's Publishing
3195 Wilson Drive NW
Grand Rapids, Michigan 49544

Learning About Money—grades 1–2
ISBN: 0-7682-0530-1

2 3 4 5 6 7 8 9 SNC 07 06 05 04 03 02

The Penny

This is a penny.

1 penny = 1 cent

1 penny = 1¢

Count each group of pennies and write the amount.

1.  _____ ¢

2.  _____ ¢

3.  _____ ¢

4. _____ ¢

5. _____ ¢

6.  _____ ¢

7.  _____ ¢

8.  _____ ¢

9.  _____ ¢

10.  _____ ¢

Brainwork! Draw a penny. Color it brown.

The Toy Shop

How much does each toy cost?
Count the pennies and write the amount.

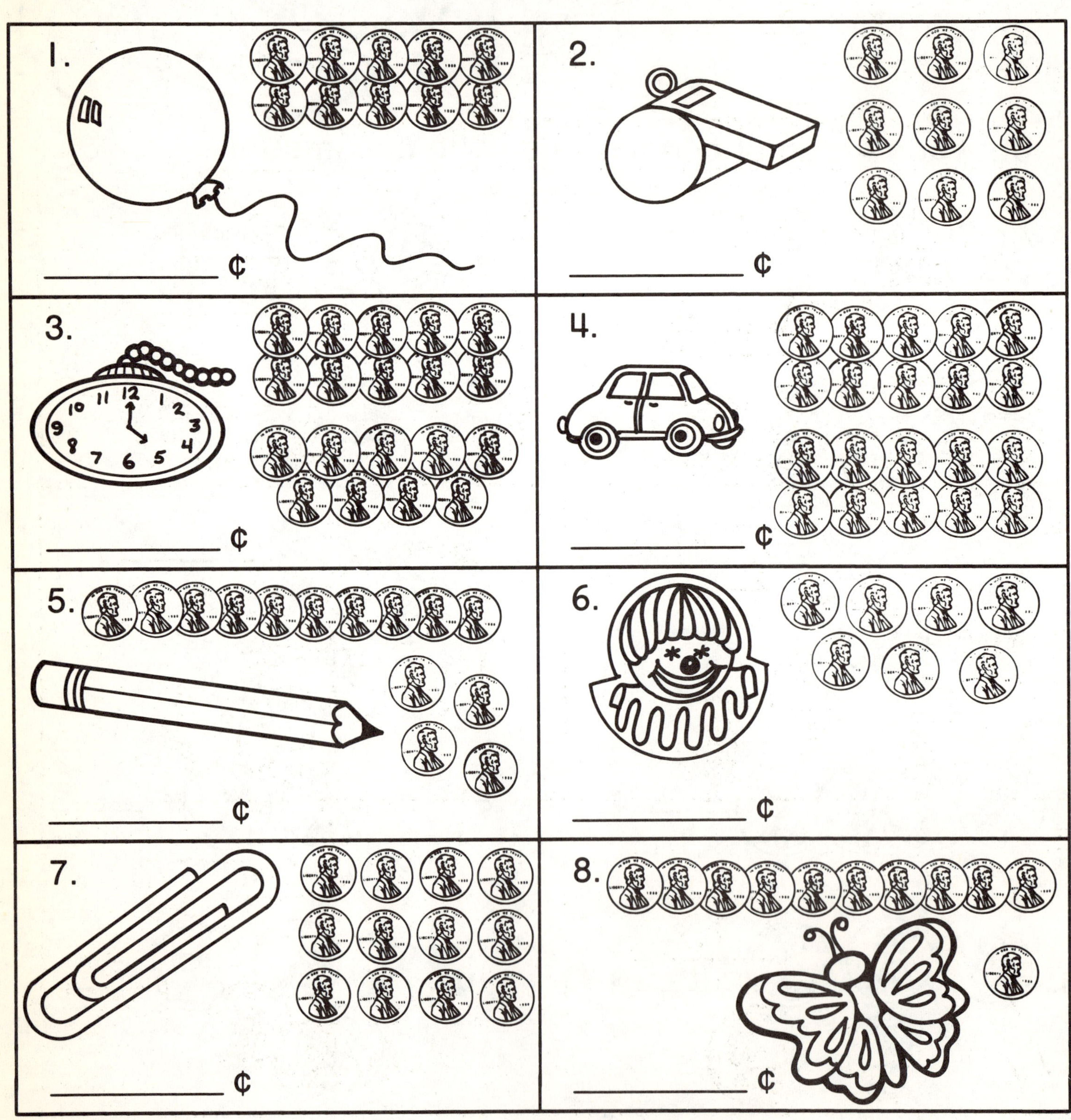

1. _______ ¢

2. _______ ¢

3. _______ ¢

4. _______ ¢

5. _______ ¢

6. _______ ¢

7. _______ ¢

8. _______ ¢

Brainwork! Draw a picture of your favorite toy.

  FS-8414 Learning About Money

What Is a Nickel?

This is a nickel.

I nickel = 5 pennies

I nickel = 5 cents

I nickel = 5¢

5¢ = 1¢ + 1¢ + 1¢ + 1¢ + 1¢

Mark an **X** on the coins you need to buy each item.

1. 6 ¢

2. 9 ¢

3. 7 ¢

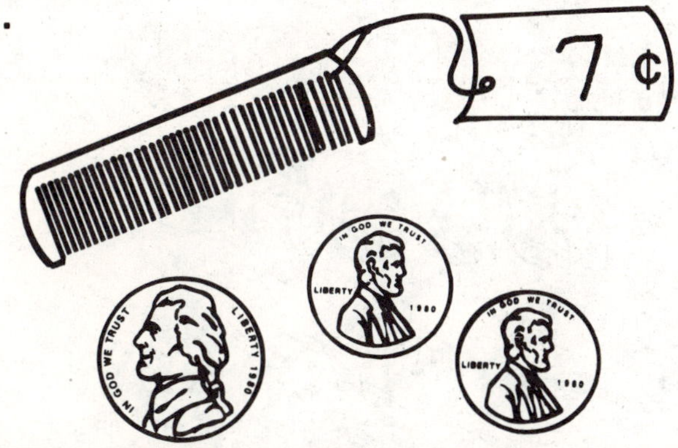

4.  5 ¢

5. 10 ¢

6. 8 ¢

Brainwork! Draw one nickel and five pennies. Write the amount.

 FS-8414   Learning About Money

Let's Count Nickels

 = 5¢

Count each group of nickels and write the amount.

1.  _____ ¢

2. _____ ¢

3.  _____ ¢

4. _____ ¢

5.  _____ ¢

6.   _____ ¢

7.  _____ ¢

8.  _____ ¢

9.  _____ ¢

10.   _____ ¢

Brainwork! Draw enough nickels and pennies to make 26¢.

          FS-8414   Learning About Money

Counting Nickels and Pennies

How much does each item cost?
Count the coins. Write the amount on the tag.

1.

2.

3.

4.

5.

6.

7.

8.

Brainwork! Color only the fruits. Circle the one you like best.

     FS-8414   Learning About Money

What Is a Dime?

This is a dime.

1 dime = 10 pennies

1 dime = 10 cents   

1 dime = 10¢

Match each set of dimes to the correct amount.

1.   ◯ ◯   10¢

2.   ◯ ◯ ◯ ◯ ◯ ◯   20¢

3.   ◯ ◯ ◯ ◯ ◯   30¢

4.   ◯ ◯ ◯ ◯ ◯ ◯ ◯ ◯   40¢

5.   ◯ ◯ ◯   50¢

6.         60¢

7.          70¢

8.           80¢

9.   ◯   90¢

Brainwork! Draw something you can buy for a dime or ten pennies.

Dimes and Pennies

Count each group of coins and write the amount.
Then solve the puzzle.

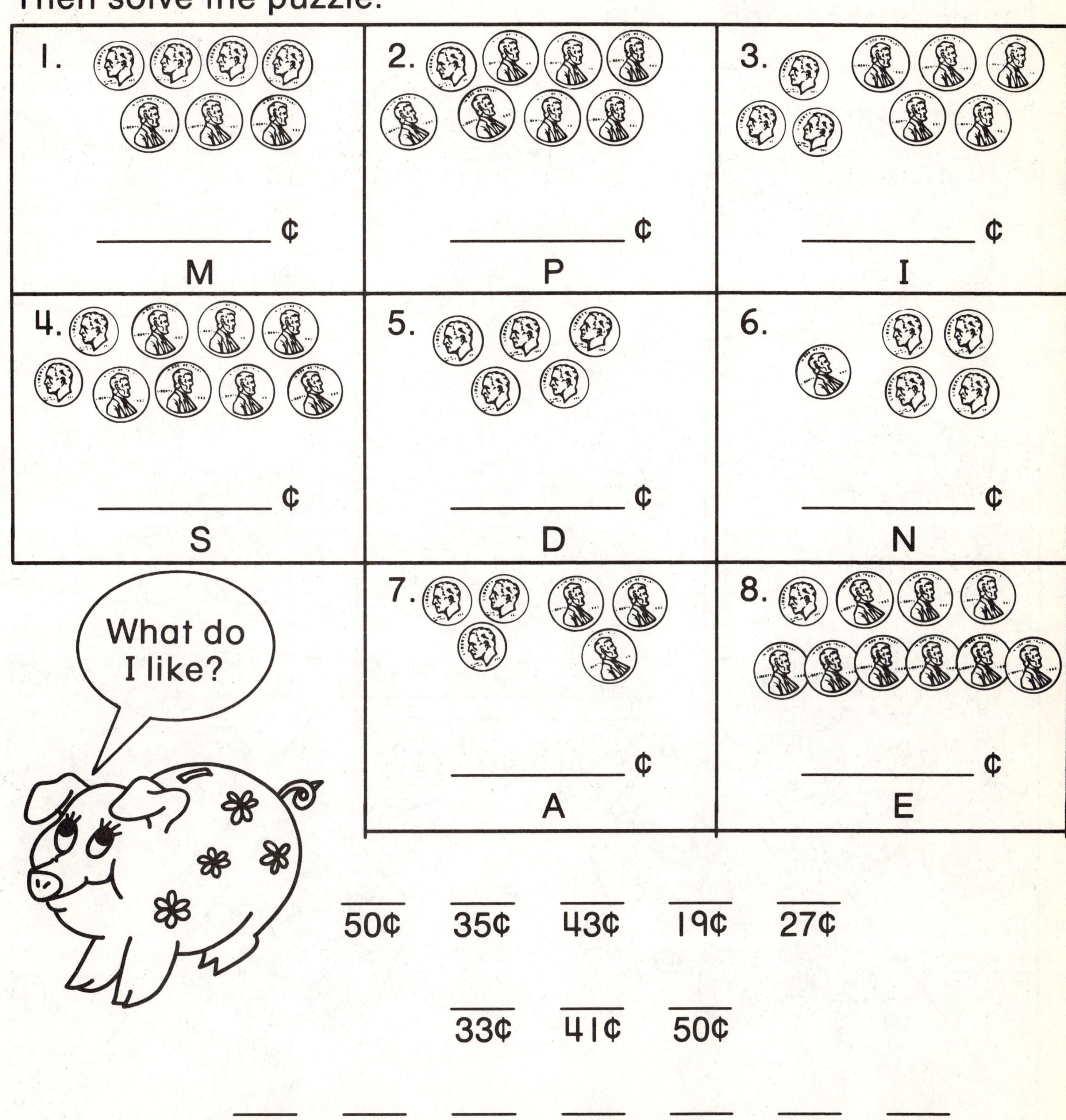

Brainwork! Draw enough dimes and pennies to equal 46¢.

The Money Man

Count each group of coins and write the amount.
Then color the man.

Brainwork! Write a story about Money Man.

  FS-8414 Learning About Money

 Dimes and Nickels 

Mark an **X** on the coins you need to buy each item.

1.  25¢

2.  45¢

3.  50¢

4.  30¢

5.  75¢

6.  40¢

7. 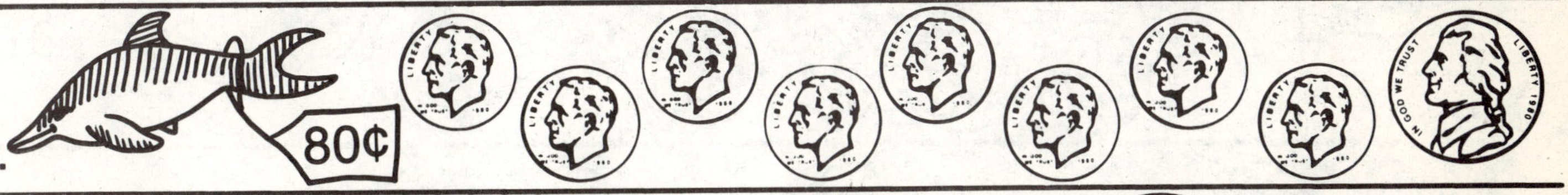 80¢

8.  65¢

9.  95¢

Brainwork! Color the animals that can swim. Circle the one you like best.

9

                    FS-8414 Learning About Money

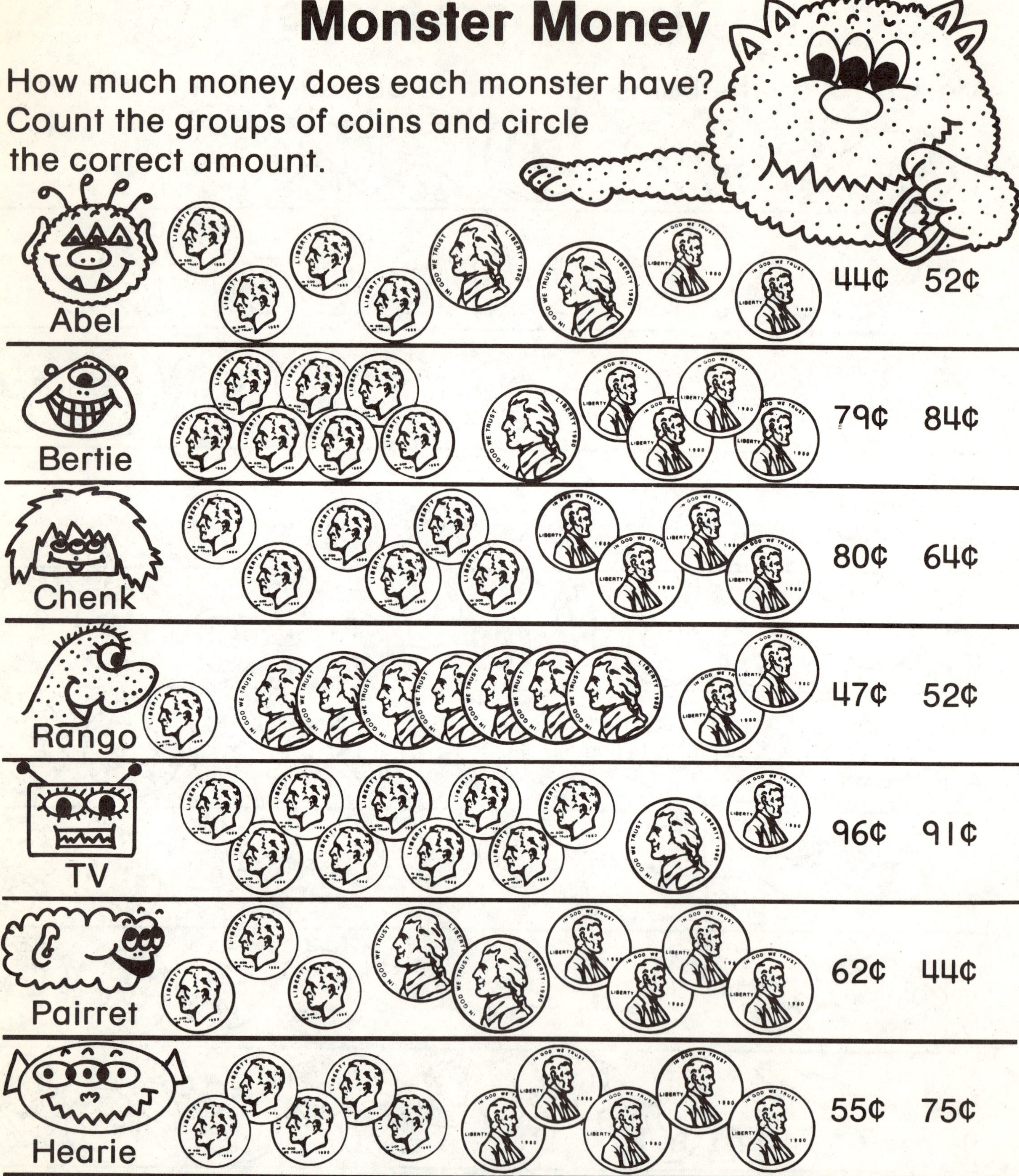

Monster Money

Name ___________________

How much money does each monster have?
Count the groups of coins and circle
the correct amount.

Brainwork! Draw a monster. Give him some coins. Count them and write the amount.

Let's Count Coins

Count the coins. Write the amount on the tag.

1. 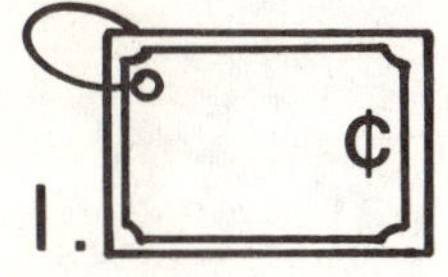  

2. 

3. 

4. 

5. 

6. 

Brainwork! Make a price tag for something that costs one dime, two nickels and three pennies.

Let's Go Shopping!

Draw the coins you need to buy each item.

d = dime
10¢

n = nickel
5¢

p = penny
1¢

 17¢    

 50¢

 63¢

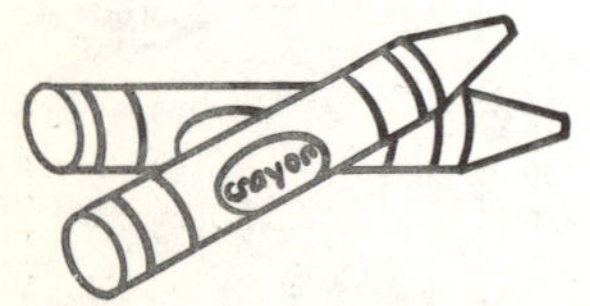 89¢

 76¢

 97¢

Brainwork! If you had only nickels, how many would you need
to buy the top?

     FS-8414   Learning About Money

Quarters

This is a quarter.

I quarter = 25 pennies

I quarter = 25 cents

I quarter = 25 ¢

10¢ + 10¢ + 5¢ = 25¢

Count each group of coins. Does it equal a quarter? Write **yes** or **no**.

1.  ~~yes~~	2.  _______
3.  _______	4.  _______
5.  _______	6.  _______
7.  _______	8.  _______
9.  _______	10.  _______

Brainwork! Draw a green circle around all the sets that equal a quarter.

    FS-8414   Learning About Money

Match It!

Match each set of coins to the correct amount.

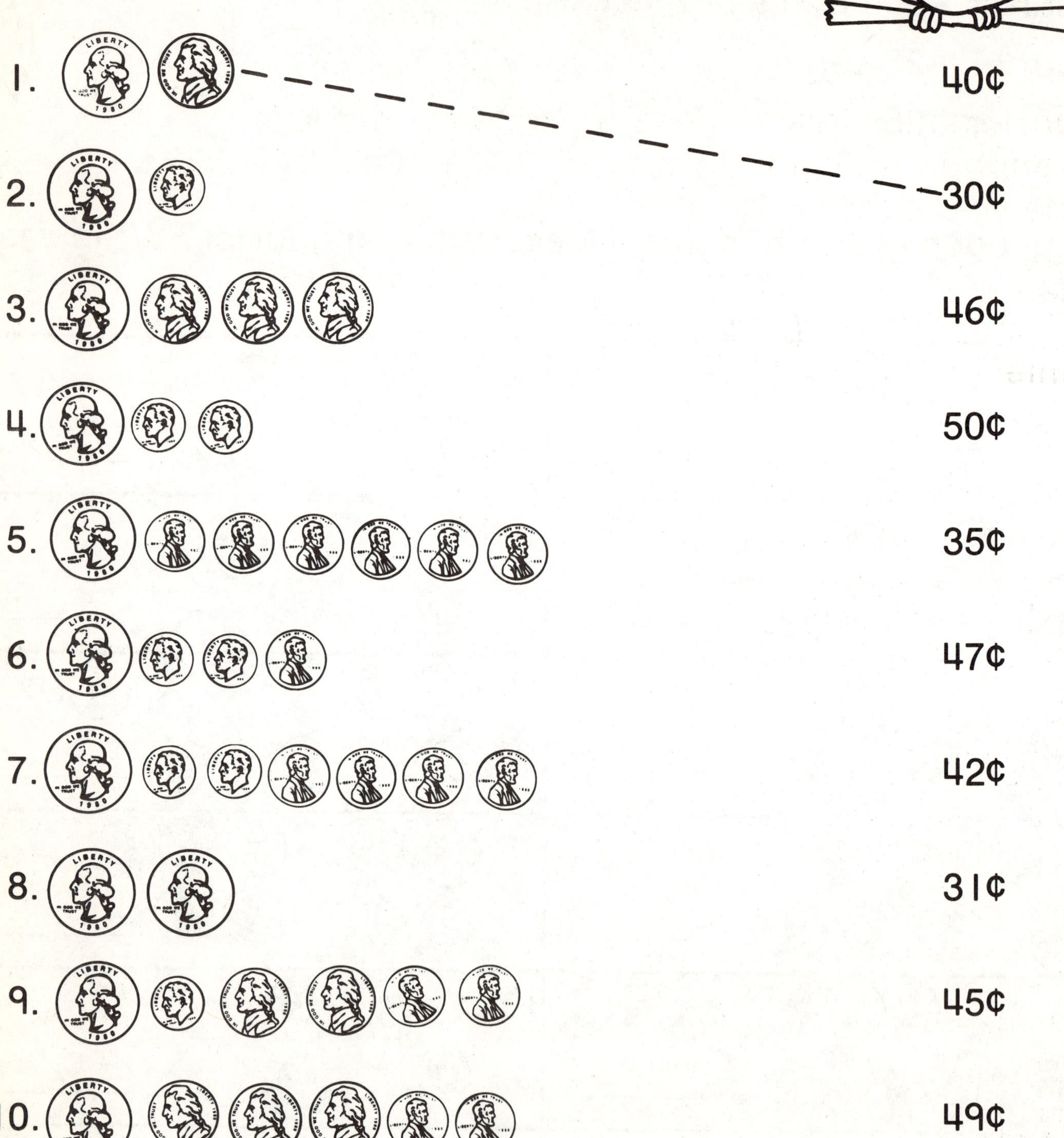

1.   40¢

2.   30¢

3.   46¢

4.   50¢

5.   35¢

6.   47¢

7.   42¢

8.   31¢

9.   45¢

10.   49¢

Brainwork! Write the amounts in order from smallest to largest.

Name ______________________________ Skill: Counting quarters, dimes, and nickels

Shopping Spree

Count each child's money and write the amount.
Is it enough to buy the pencil?
Circle yes or no.

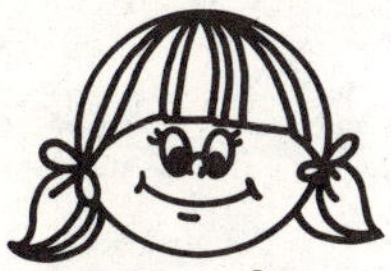
Annie
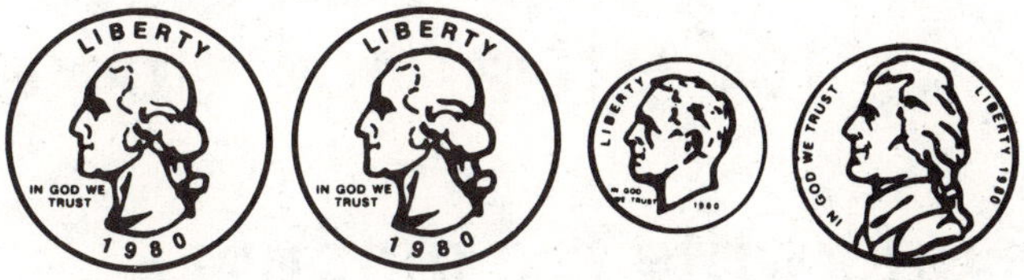

_____ ¢ yes no

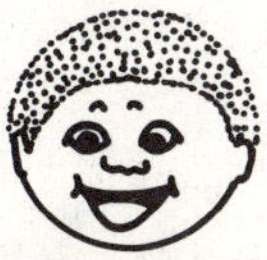
Sam
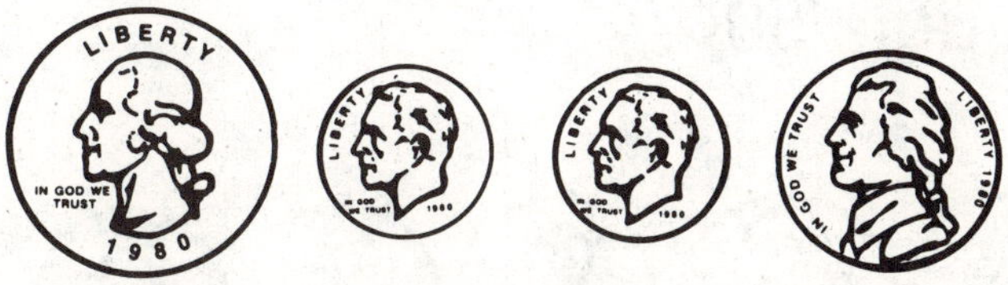

_____ ¢ yes no

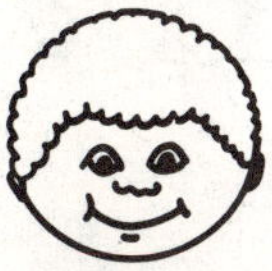
Marc
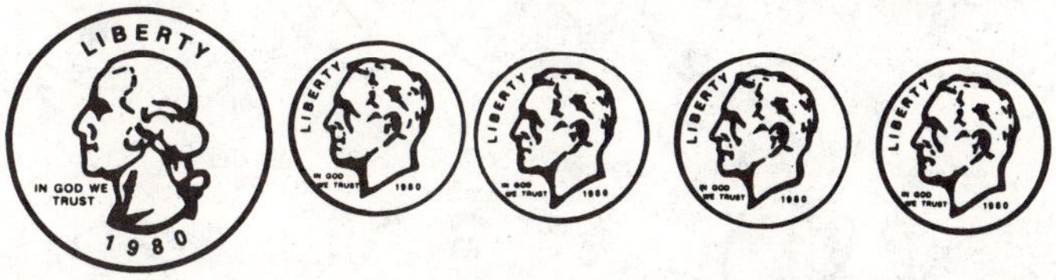

_____ ¢ yes no

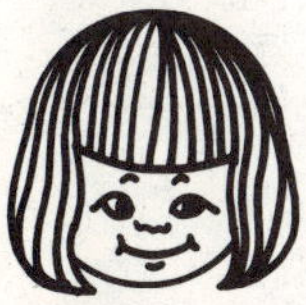
Mary
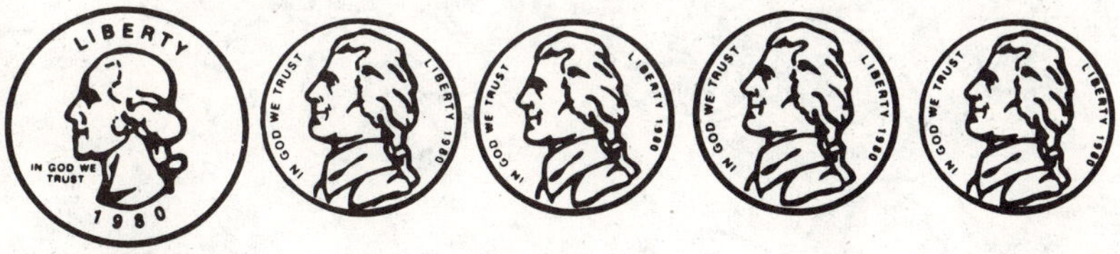

_____ ¢ yes no

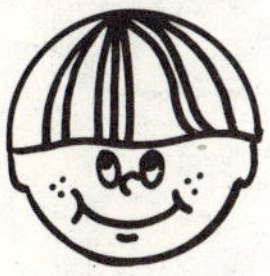
Bob
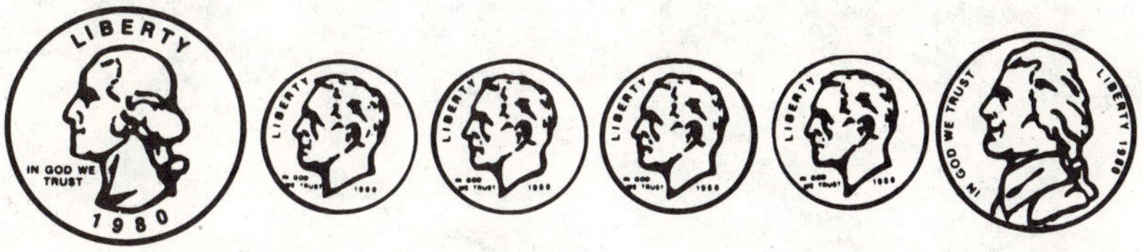

_____ ¢ yes no

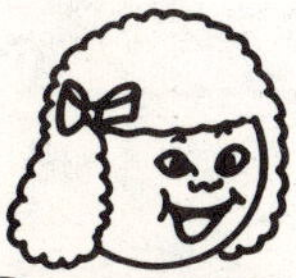
Carmen
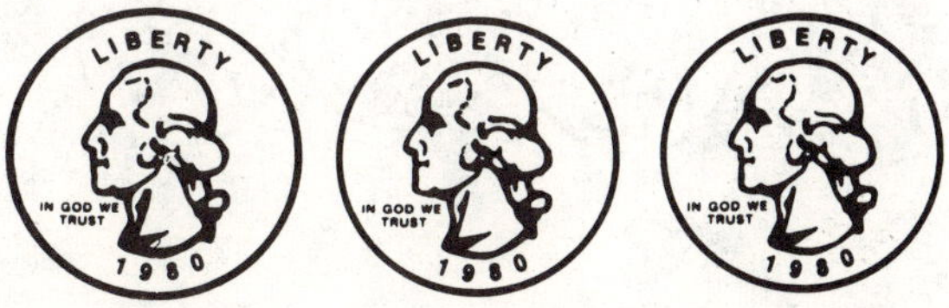

_____ ¢ yes no

Brainwork! How much money do Sam and Mary have together?

Name ______________________________

Skill: Half dollar recognition,
Counting sets of coins

Half Dollars

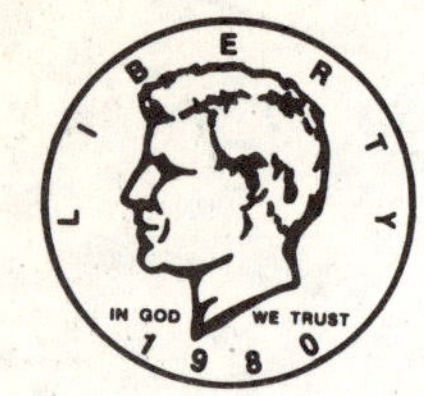

This is a half dollar.

I half dollar = 50 pennies

I half dollar = 50 cents

I half dollar = 50¢

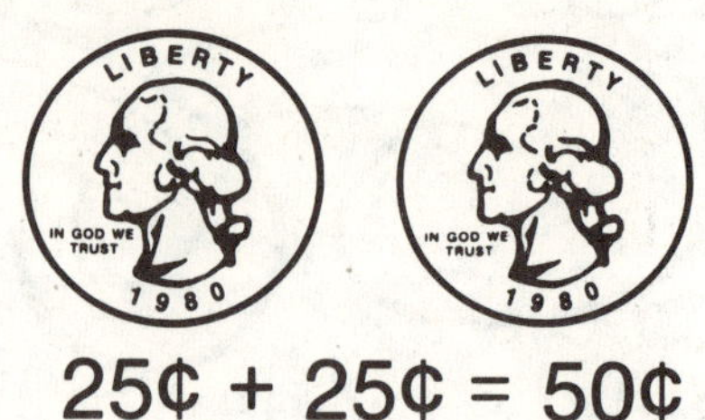

25¢ + 25¢ = 50¢

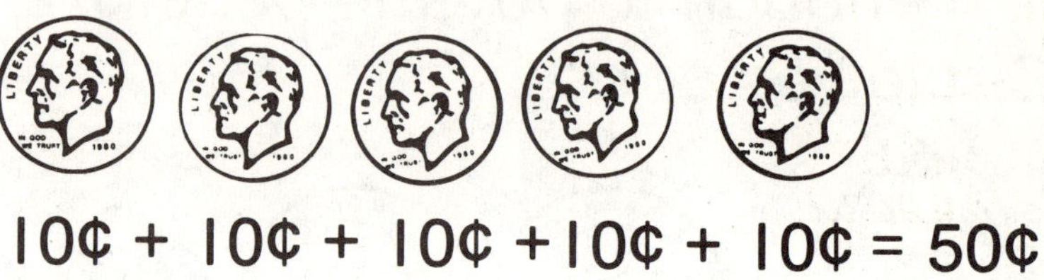

10¢ + 10¢ + 10¢ +10¢ + 10¢ = 50¢

Count each group of coins and write the amount.
Remember! A half dollar equals 50¢.

1. 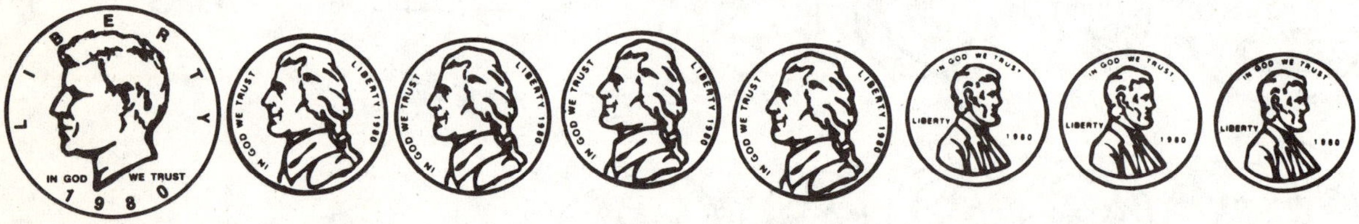 ______ ¢

2. 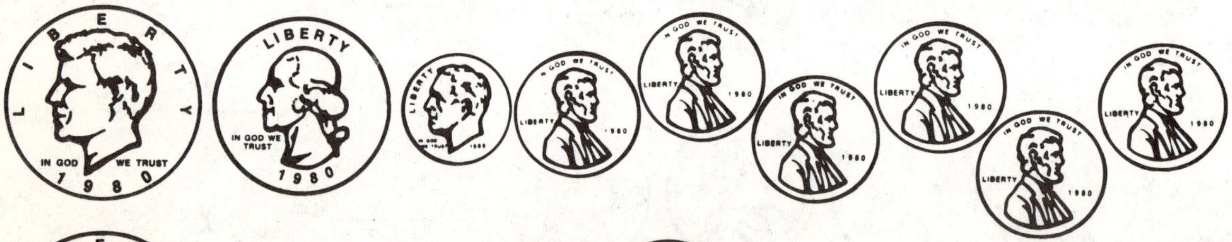 ______ ¢

3. 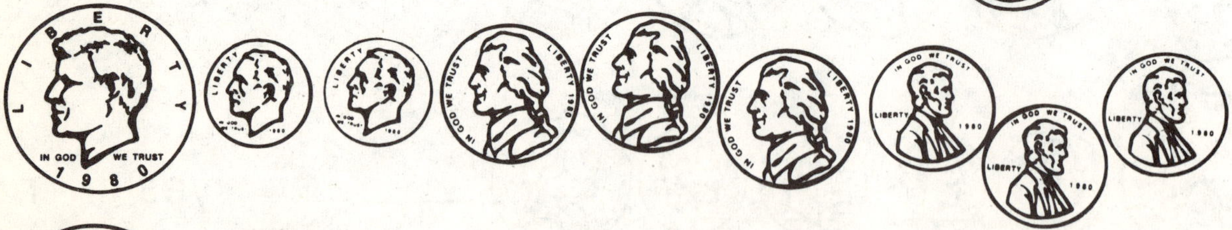 ______ ¢

4. 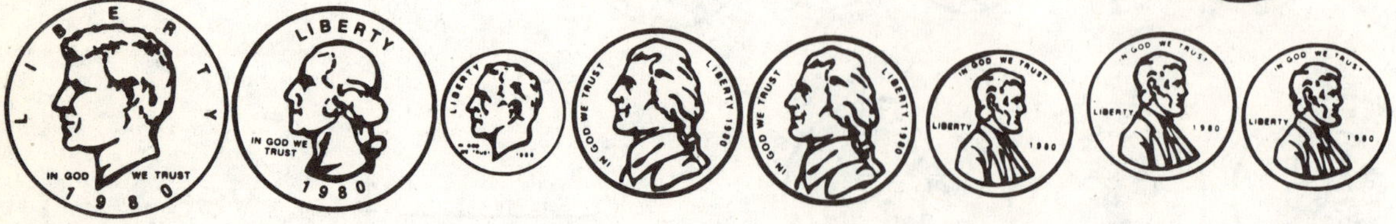 ______ ¢

Brainwork! Draw the coins you need to buy lunch at your school. Write the amount.

Which Coin Is Missing?

Count each set of coins. Then look at the price tag.
One more coin is needed to equal the amount on the price tag.
Find the missing coin at the bottom of the page. Cut and paste it
to complete each set.

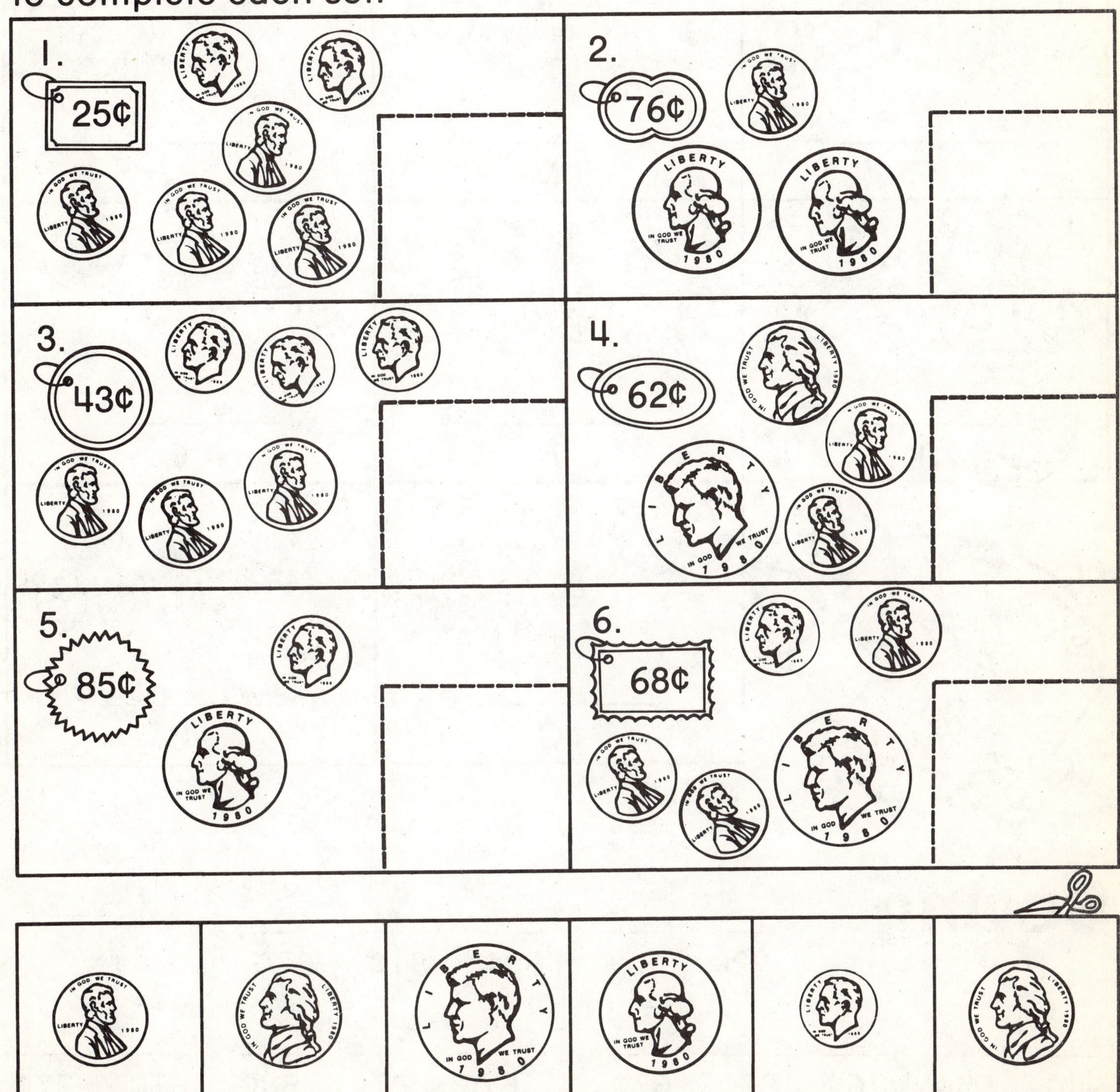

FS-8414   Learning About Money

A Puzzle Page

Count each set of coins and write the amount.
Then solve the puzzle.

1. _______ ¢ T	2. _______ ¢ M	3. _______ ¢ N
4. _______ ¢ K	5. _______ ¢ A	6. _______ ¢ E
	7. _______ ¢ C	8. _______ ¢ S

Saving

___ ___ ___ ___ ___
92¢ 87¢ 88¢ 81¢ 67¢

___ ___ ___ ___ ___ ___ ___ ___ ___ ___ .
94¢ 59¢ 76¢ 87¢ 67¢ 67¢ 87¢ 88¢ 67¢ 87¢

18

Name ______________________________

Skill: Dollar recognition,
Counting coins, Money signs

The Dollar

This is a dollar.

I dollar = 100 pennies

I dollar = 100 cents

I dollar = $1.00

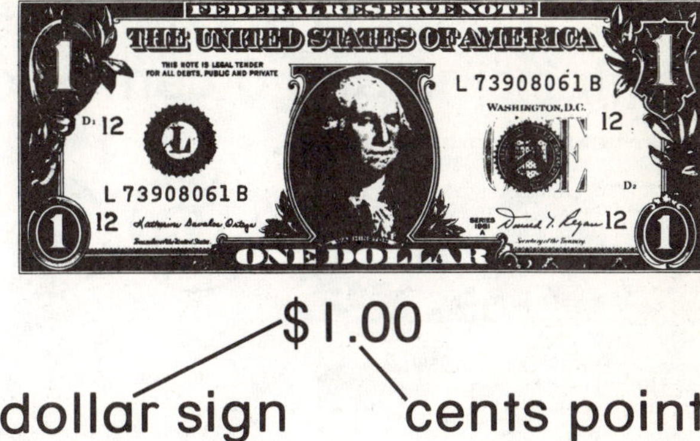

$1.00

dollar sign — cents point

Count each set of coins and write the amount.
If it equals one dollar, use the dollar sign and cents point in your answer. **($1.00)**

1. 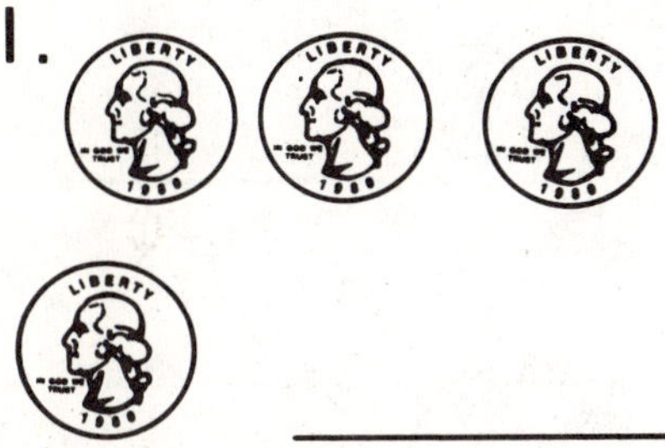 ______	2. 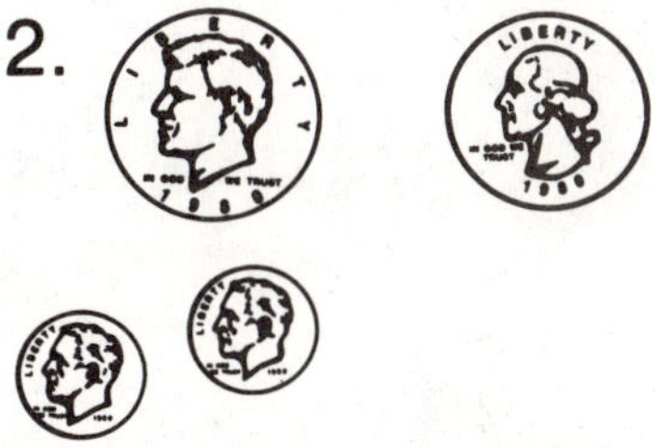 ______	3. 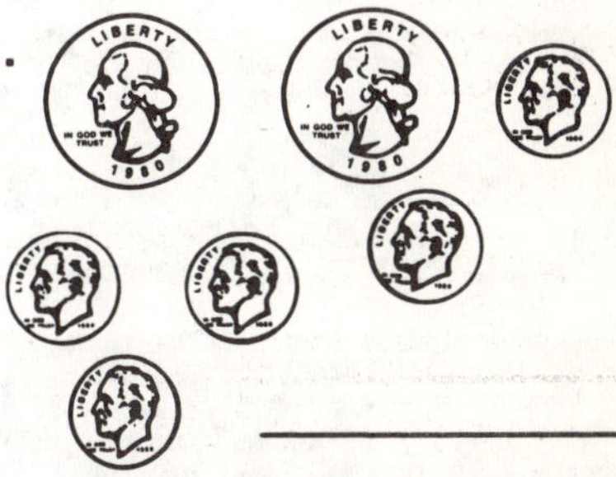 ______
4. 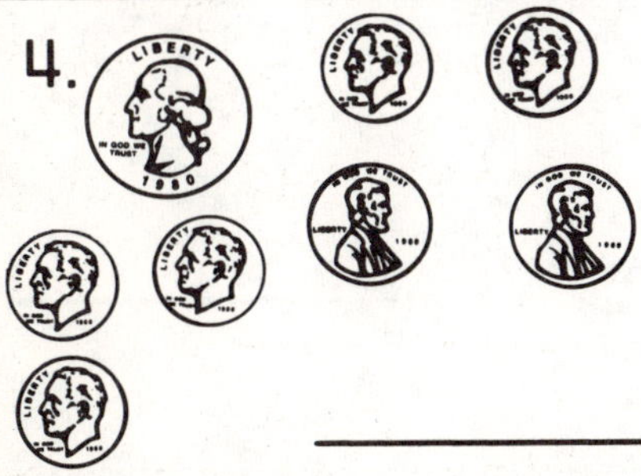 ______	5. 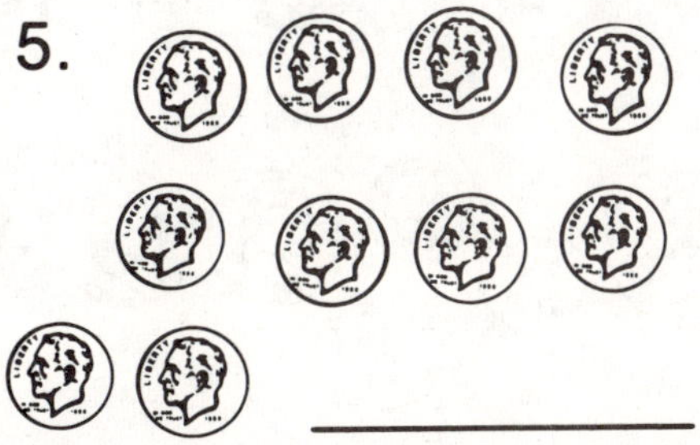 ______	6. 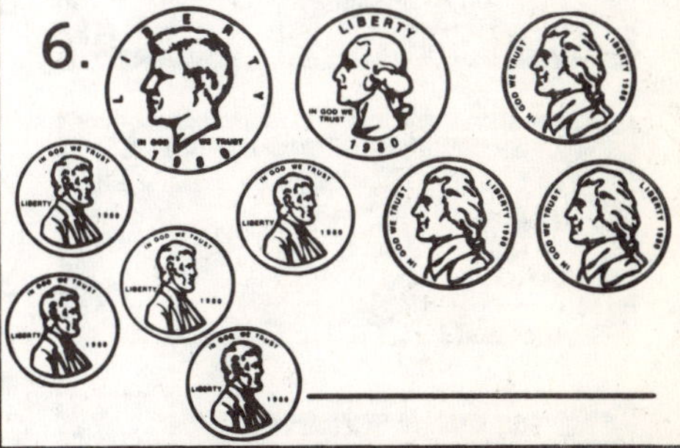 ______
7. 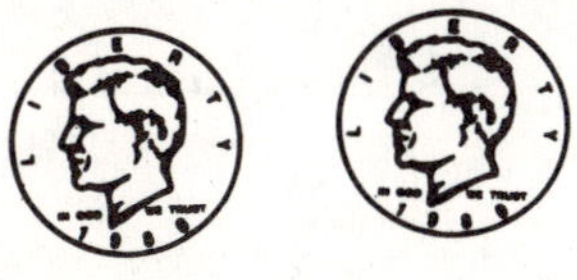 ______	8. 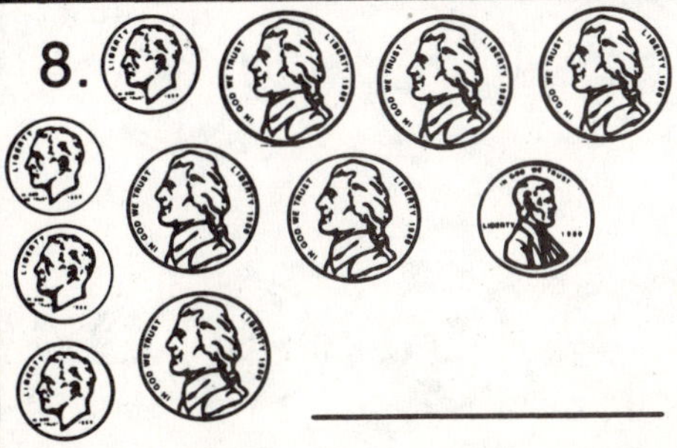 ______	9. 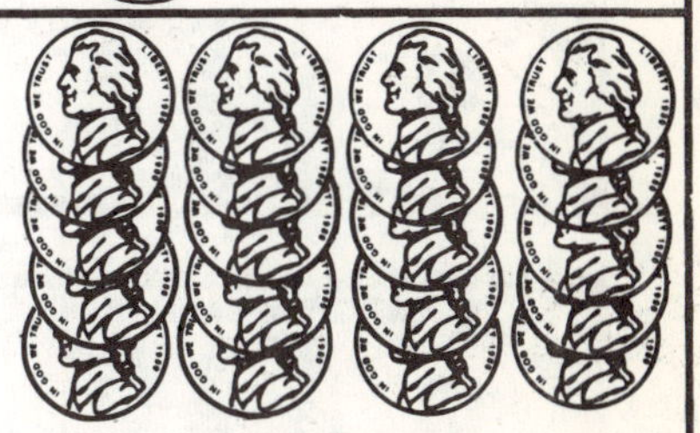 ______

Brainwork! Color all the boxes that equal $1.00 **purple**.

Dollars and Cents

How much money is in each set?
Use a dollar sign and a cents point as you write the amounts.

1.   
$1.63

2.   ____________

3.   ____________

4.   ____________

5.   ____________

6.   ____________

7.   ____________

8.   ____________

Brainwork! Write the amounts in order from the lowest to the highest in value.

Answers

Page One

1. 3¢
2. 4¢
3. 5¢
4. 2¢
5. 1¢
6. 5¢
7. 4¢
8. 2¢
9. 4¢
10. 5¢

Page Two

1. 10¢
2. 9¢
3. 19¢
4. 20¢
5. 14¢
6. 7¢
7. 12¢
8. 11¢

Page Three

1. one nickel, one penny
2. one nickel, four pennies
3. one nickel, two pennies
4. one nickel
5. two nickels
6. one nickel, three pennies

Page Four

1. 15¢
2. 30¢
3. 50¢
4. 25¢
5. 20¢
6. 50¢
7. 10¢
8. 35¢
9. 40¢
10. 45¢

Page Five

1. 23¢
2. 35¢
3. 19¢
4. 50¢
5. 32¢
6. 27¢
7. 14¢
8. 28¢

Page Six

1. 20¢
2. 60¢
3. 50¢
4. 80¢
5. 30¢
6. 70¢
7. 40¢
8. 90¢
9. 10¢

Page Seven

1. 43¢
2. 17¢
3. 35¢
4. 27¢
5. 50¢
6. 41¢
7. 33¢
8. 19¢

Solution: dimes and pennies

Page Eight

1. 62¢
2. 35¢
3. 44¢
4. 14¢
5. 85¢
6. 70¢
7. 54¢
8. 18¢
9. 38¢
10. 92¢

Page Nine

1. one dime, three nickels
2. four dimes, one nickel
3. four dimes, two nickels
4. three dimes
5. seven dimes, one nickel
6. one dime, six nickels
7. eight dimes
8. five dimes, three nickels
9. nine dimes, one nickel

Page Ten

Abel	52¢
Bertie	79¢
Chenk	64¢
Rango	47¢
TV	96¢
Pairret	44¢
Hearie	55¢

Page Eleven

1. 74¢
2. 49¢
3. 46¢
4. 97¢
5. 91¢
6. 77¢

Page Twelve

Answers will vary.

Page Thirteen

1. yes
2. no
3. no
4. yes
5. yes
6. no
7. yes
8. yes
9. yes
10. no

Page Fourteen

1. 30¢
2. 35¢
3. 40¢
4. 45¢
5. 31¢
6. 46¢
7. 49¢
8. 50¢
9. 47¢
10. 42¢

Page Fifteen

Annie	65¢	yes
Sam	50¢	no
Marc	65¢	yes
Mary	45¢	no
Bob	70¢	yes
Carmen	75¢	yes

Page Sixteen

1. 73¢
2. 91¢
3. 88¢
4. 98¢

Page Seventeen

1. penny
2. quarter
3. dime
4. nickel
5. half dollar
6. nickel

Page Eighteen

1. 81¢
2. 94¢
3. 88¢
4. 76¢
5. 59¢
6. 87¢
7. 92¢
8. 67¢

Solution: cents makes sense

Page Nineteen

1. $1.00
2. 95¢
3. $1.00
4. 77¢
5. $1.00
6. 95¢
7. $1.00
8. 71¢
9. $1.00

Page Twenty

1. $1.63
2. $1.15
3. $1.68
4. $1.44
5. $1.37
6. $1.58
7. $1.65
8. $1.84